I0756103

THIS BOOK BELONGS TO:

THE WONDERFUL WORLD OF DUCKS
MIMI JONES

Dedicated to all who love ducks.

ISBN 978-1-970416-07-7

www.joeysavestheday.com

Mimi Books™ Publishing

A Mimi Book

Ducks belong to a group of water birds called waterfowl, along with geese and swans.

There are more than 120 different species of ducks around the world.

GARGANEY

COMMON POCHARD

AFRICAN BLACK DUCK

MALLARD

WHITE-CHEEKED PINTAIL

MOTTLED DUCK

Ducks can live in freshwater, saltwater, and even brackish water, which is a mix of both.
FRESHWATER
SALTWATER
BRACKISH WATER

Ducks are found on every continent except Antarctica.

WATERPROOF

Ducks have waterproof feathers that help them stay warm and dry in the water.

Ducks have webbed feet that act like paddles, helping them swim quickly.

A duck's bill is soft and flexible, perfect for scooping food from the water. Their bills have tiny comb-like edges called lamellae that help them filter food.

Ducks have excellent vision and can see almost 360 degrees around them.

Ducks have strong wings that help them fly long distances during migration. Some ducks migrate thousands of miles every year to find warmer weather and food.

MIGRATION

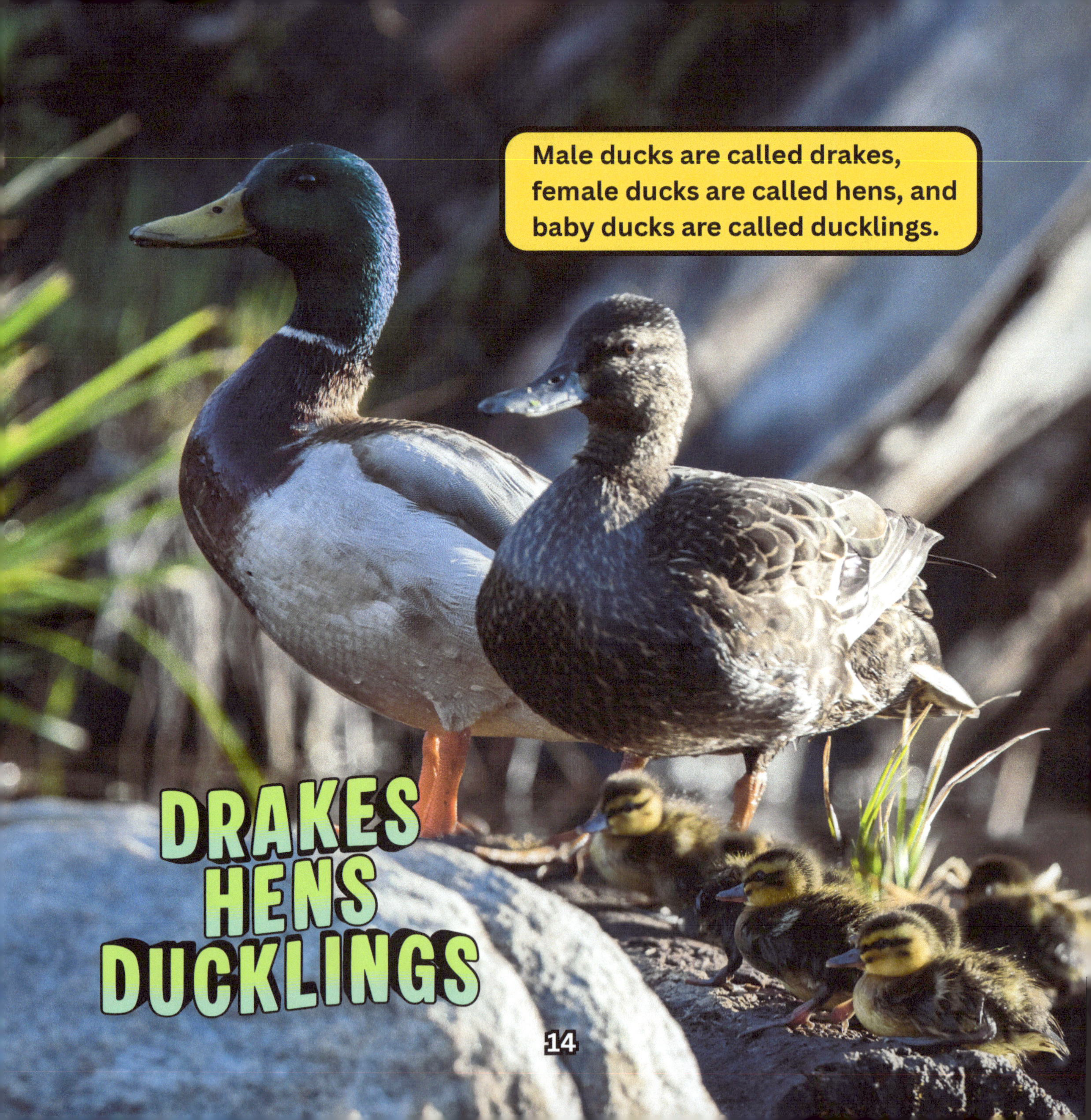
Male ducks are called drakes, female ducks are called hens, and baby ducks are called ducklings.
DRAKES
HENS
DUCKLINGS

Male ducks are usually more colorful than females.
COLORFUL

Ducks are omnivores, which means they eat plants, seeds, insects, and even small animals like fish.
OMNIVORE

Ducks often tip forward in the water to reach food underwater; this is called dabbling.
DABBLING

Some ducks dive deep underwater to catch fish and plants; these are called diving ducks.

DIVING

Ducks can sleep on land or on water, depending on where they feel safest. They also can sleep with one eye open to stay alert for predators.

SLEEPING

Ducks sometimes play by splashing, chasing, and diving with other ducks.

PLAYING

FLOCKS

Ducks are social animals and enjoy being around other ducks. They often travel in groups called flocks for protection.

Ducks live in ponds, lakes, rivers, marshes, and wetlands. Wetlands are important because they give ducks food, shelter, and safe nesting spots.

Ducks use the stars, the sun, and Earth's magnetic field to help them navigate during migration.

NAVIGATE

Many ducks return to the same nesting area year after year.

NESTING

Female ducks build nests out of grass, leaves, and soft feathers.

A mother duck lays a clutch of eggs, usually between 6 and 12 at a time. The eggs take about 28 days to hatch.
6-12

SWIM & WALK

Ducklings can walk, swim, and follow their mother within hours of hatching. They stay close to their mother for protection and warmth.

A group of ducklings following their mother is called a brood.

BROOD

SAFE

Mother ducks teach their ducklings where to find food and how to stay safe.

Ducks can imprint on the first moving thing they see after hatching, sometimes even humans.
IMPRINT

Ducklings grow new feathers called juvenile feathers before learning to fly. Most learn to fly when they are about 6 to 8 weeks old.

Ducks communicate with quacks, whistles, grunts, and soft peeps. They also use body language, like head bobbing, to talk to each other.

Ducks can swim, walk, and fly, making them excellent all-around movers.

EXCELLENT

ICY

Ducks can stay warm in icy water because their feet have very few nerves and blood vessels.

Ducks can dive underwater for up to 30 seconds, depending on the species.

UP TO 30 SECONDS

Some ducks can fly up to 60 miles per hour during migration.

60 MILES PER HOUR

Count the ducks.
Thank you for exploring The Wonderful World of Ducks with me. I hope you learned something new and had fun discovering these curious, quacky, feathered friends
If you enjoyed this book, please consider leaving a review. It helps other families discover it too.
See you in the next adventure!

Check out these other interesting books in the Wonderful World of series!

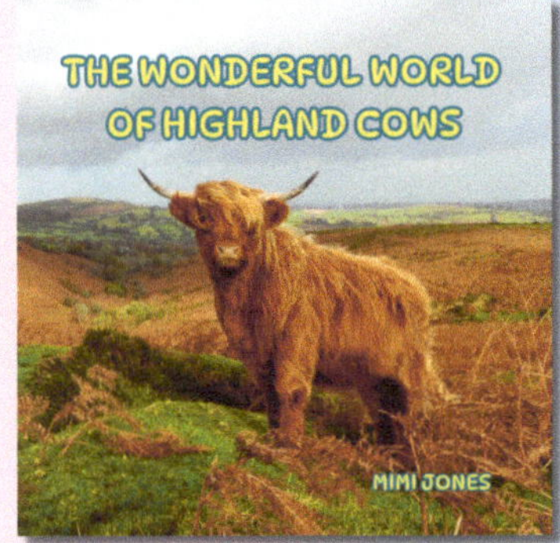

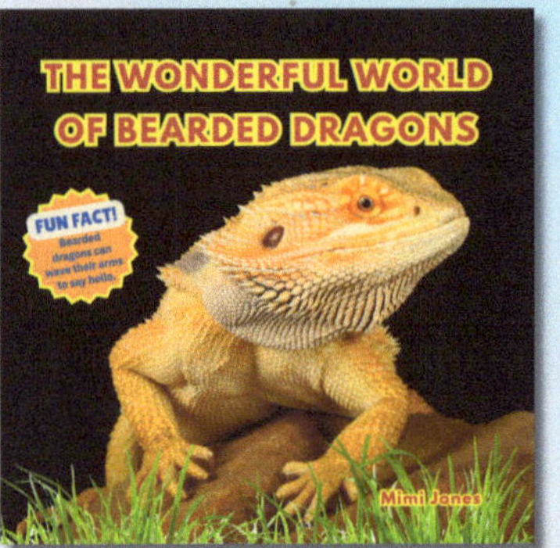

www.mimibooks.com

www.ingramcontent.com/pod-product-compliance
Lightning Source LLC
LaVergne TN
LVHW070158110826
845147LV00002B/436

9781970416077